AF575881

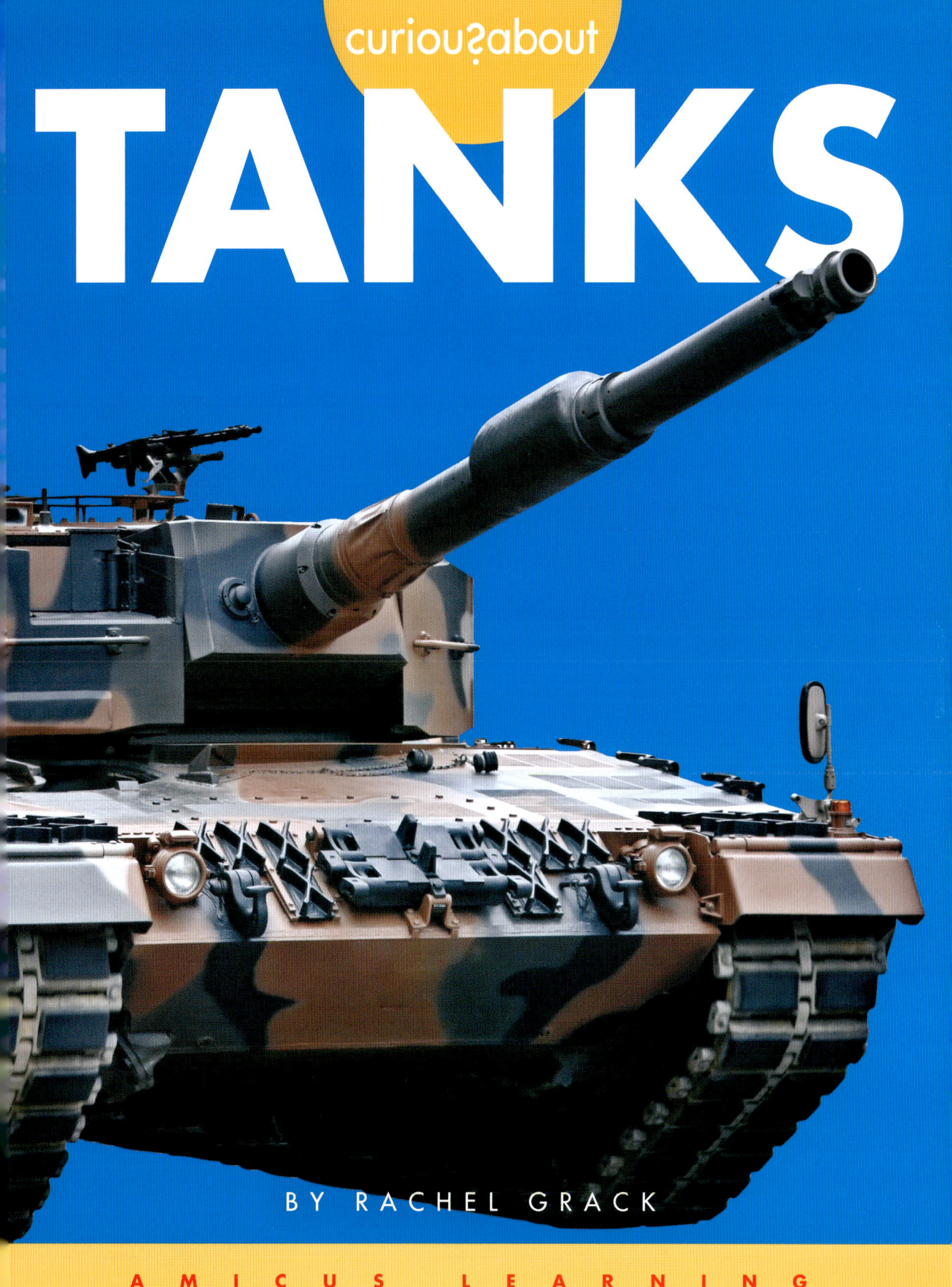
curious about
TANKS
BY RACHEL GRACK
AMICUS LEARNING

What are you

curious about?

CHAPTER THREE

Curious About is published by
Amicus Learning, an imprint of Amicus
P.O. Box 227
Mankato, MN 56002
www.amicuspublishing.us

Editor: Alissa Thielges
Series Designer: Kathleen Petelinsek
Book Designer: Aubrey Harper

Library of Congress Cataloging-in-Publication Data
Names: Koestler-Grack, Rachel A., 1973– author.
Title: Curious about tanks / by Rachel Grack.
Description: Mankato, MN : Amicus Learning, [2025] | Series: Curious about military machines | Includes bibliographical references and index. | Audience: Ages 5–9 | Audience: Grades 2–3 | Summary: "Early elementary readers learn how tanks work in this inquiry-based nonfiction book about their size, speed, and battle strength. Includes infographics and back matter to support research skills, plus table of contents, glossary, and index"—Provided by publisher.
Identifiers: LCCN 2023038620 (print) | LCCN 2023038621 (ebook) | ISBN 9781645493198 (library binding) | ISBN 9781645494072 (ebook)
Subjects: LCSH: Tanks (Military science)—Juvenile literature. | Armored vehicles, Military—Juvenile literature.
Classification: LCC UG446.5 .K6132 2025 (print) | LCC UG446.5 (ebook) | DDC 623.74/752—dc23/eng/20231107
LC record available at https://lccn.loc.gov/2023038620
LC ebook record available at https://lccn.loc.gov/2023038621

Photo credits: Alamy/INTERFOTO/History 21 (2), Niday Picture Library 21 (4); Department of Defense 4–5; DVIDS/Capt. Thomas Barger 11 (IFV), Cpl. Alisha Grezlik 14–15, Cpl. Gabrielle Quire 17, Lance Cpl. Menelik Collins 12–13, Marine Corps Sgt. M. Trent Lowry 10, Sgt. Alexis Flores 20, Sgt. Chad Menegay 18–19, Sgt. Leon Cook 9, Sgt. Matthew Lucibello 11 (APC); iStock/huettenhoelscher 8, vasiliki cover, 1, 11 (tank); Noun Project/Sandhi Priyasmoro 22 & 23 (icons); Shutterstock/ID1974 16; Wikimedia Commons/German Federal Archive 21 (3), HMSO 6–7, Mil.ru 21 (1), PHC HOLMES 21 (5)

Printed in China

CHAPTER ONE

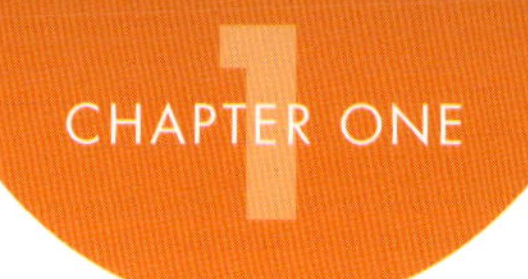

What are tanks?

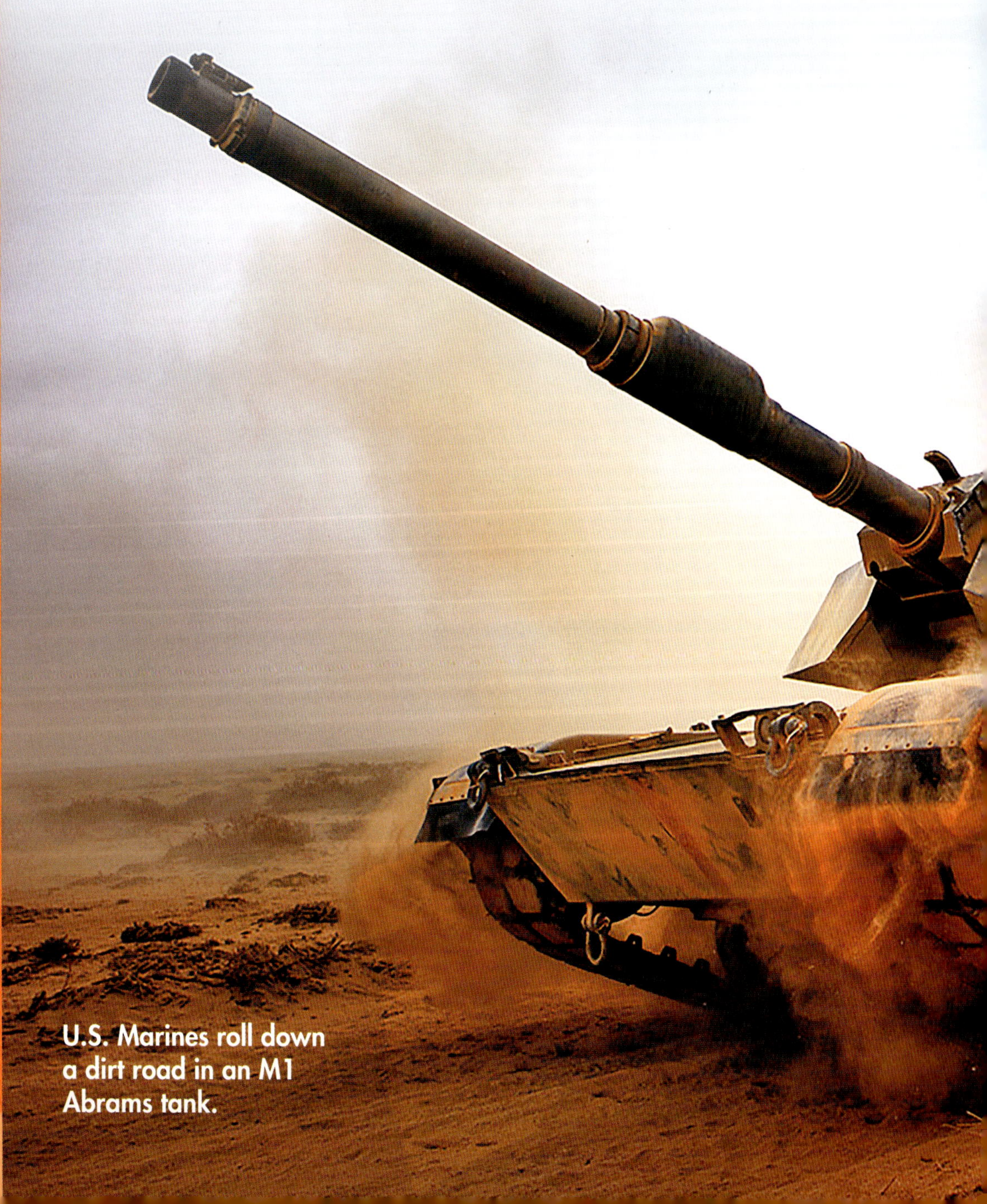

U.S. Marines roll down a dirt road in an M1 Abrams tank.

Tanks are **armored** fighting vehicles (AFVs). They blast through enemy lines. Bullets bounce off the **hull**. Wide tracks wrap around the wheels. They roll over **obstacles** and uneven ground. Their large cannon and machine guns turn in any direction. Ready. Aim. Fire!

The very first tank was named "Little Willie."

Who built the first tank?

Great Britain did in World War I (1914–18). Enemies hid in **trenches** behind barbed wire. Crossing a trench was dangerous and deadly. But there was no other way around them. The British built an AFV to carry soldiers safely across. At first, they kept the vehicle a secret. They called it a "water tank." The "tank" part stuck.

CHAPTER TWO

How powerful are tanks?

M1 Abrams produce 1,500 horsepower. That's the same as two race cars. M1s are the U.S. Army's main battle tanks. Their engines are powerful yet lightweight. M1s can **accelerate** faster than any other tanks. They go from 0 to 20 miles (32 kilometers) per hour in 7.2 seconds.

M1 ABRAMS TANK

Top Speed: 42 miles (68 km) per hour
Weight: 70 tons (64 metric tons)
Weapons: 1 cannon, 3 machine guns
Vertical Obstacle: 42 inches (107 centimeters)
Max Width of Trench Crossing: 9 feet (2.7 m)

An M1 tank powers through barbed wire and sand during a training exercise.

What fuel do tanks use?

Both trucks and helicopters can help tanks refuel.

Tanks use gasoline, diesel, and jet **fuel**. M1 tanks hold 490 gallons (1,850 liters). That sounds like a lot. But they can only travel 265 miles (426 km) on that amount. They get less than 1 mile per gallon (0.4 km/liter)!

Infantry Fighting Vehicle (IFV): Carries soldiers and offers extra firepower. Heavily armed.

Armored Personnel Carrier (APC): Carries troops into battle. Lightly to moderately armed. Faster than tanks but cannot travel over rough ground.

Tank: Used for battle. Heavily armed. Armored against almost any weapon.

ARMORED FIGHTING VEHICLES

Tanks have thicker armor in the front of the hull.

What makes tanks bulletproof?

DID YOU KNOW?
If a weapon breaks through, onboard safety systems quickly put out the fire and clean the air.

The exact materials are a secret! But we know a few. The outside is covered in thick, steel plates. Inside walls are, too. Ceramic blocks get positioned between the walls. These take the heat from missiles. Air pockets catch gases and pieces of metal. Not much can get through to the crew.

CHAPTER THREE

3

How many people fit in a tank?

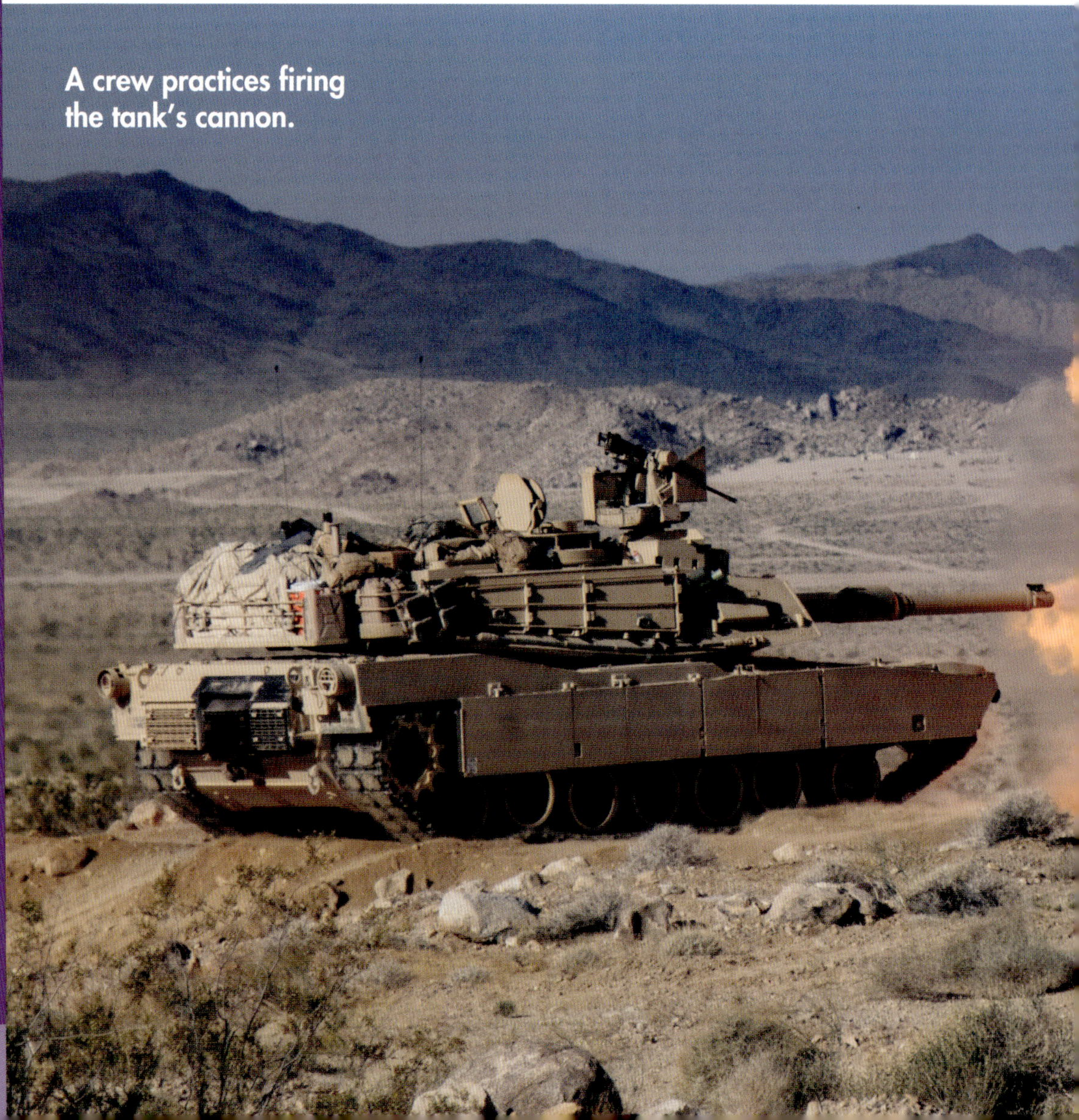

A crew practices firing the tank's cannon.

Most tanks carry a four-soldier crew. A loader puts **ammunition** into the main gun. The gunner fires at the enemy. Another solider drives the tank. The commander oversees all of the action. He directs the crew and communicates with other tanks.

DID YOU KNOW?
Someday robot tanks could roll into battle without any crew at all.

Some tanks use cameras to help the driver see.

What's it like to drive one?

Crowded and tricky. The seat is small and tips way back. There is very little space to move. Drivers steer with a motorcycle-like handlebar. Tanks have no windows. Drivers use **periscopes** to find their way. They drive through the flames and smoke of battle!

A marine listens to a headset while sitting in the driver's seat of a tank.

What was the biggest tank battle ever?

Tanks are hard to beat in a battle.

The Battle of Kursk during World War II (1939–45). As many as 10,500 tanks fought during the two-month battle. But the scariest tank fight happened in the Gulf War (1990–91). More than 3,000 tanks faced off in the Iraqi desert. The battle boomed and blazed for 36 hours. It is known as Fright Night.

5
FRIGHT NIGHT
GULF WAR, 1991
TOTAL TANKS: 3,000

4
BATTLE OF THE BULGE
WWII, 1944–45
TOTAL TANKS: 3,400

3
BATTLE OF BRODY
WWII, 1941
TOTAL TANKS: 3,850

2
BATTLE OF
BIALYSTOK-MINSK
WWII, 1941
TOTAL TANKS: 6,458

1
BATTLE OF KURSK
WWII, 1943
TOTAL TANKS:
UP TO 10,500

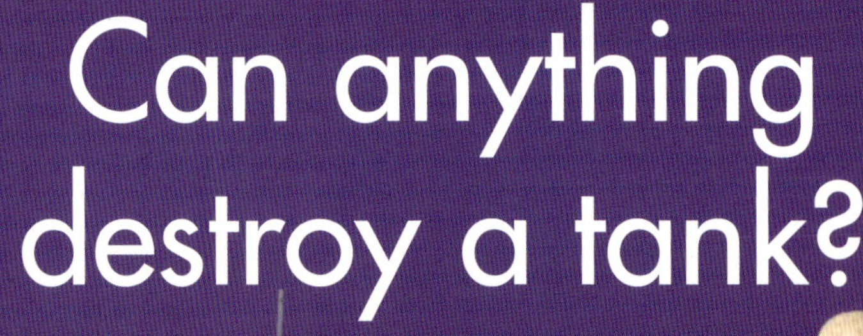

Can anything destroy a tank?

While tanks are strong, the crew is careful to avoid gunfire when looking out the hatch.

Yes. There are anti-tank missiles. Hellfire missiles are the strongest. They can destroy any tank in the world. These missiles are guided by a **laser**. They lock on to **targets** and hardly ever miss. But most other missiles won't stop a tank. Roll on!

ASK MORE QUESTIONS

Which country has the most battle tanks?

How much does a battle tank cost?

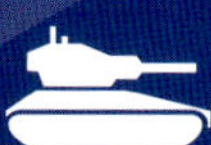

Try a BIG QUESTION: How do tanks help win a war?

SEARCH FOR ANSWERS

Search the library catalog or the Internet.
A librarian, teacher, or parent can help you.

Using Keywords
Find the looking glass.

Keywords are the most important words in your question.

If you want to know:

- who has the most tanks, type: BATTLE TANKS BY COUNTRY
- the cost of a battle tank, type: BATTLE TANK COST

FIND GOOD SOURCES

Here are some good, safe sources you can use in your research.
Your librarian can help you find more.

Books

How Tanks Work
by Walt Brody, 2020.

Tanks in Action
by Mari Bolte, 2024.

Internet Sites

DK Find Out!: Tank Warfare
dkfindout.com/us/history/world-war-i/tank-warfare
DK Find Out! is a website for kids. Discover educational topics through interactive pictures and text.

Kiddle: Tank facts for kids
https://kids.kiddle.co/Tank
Kiddle is an online encyclopedia for kids. Search this educational site for information on almost any topic!

Every effort has been made to ensure that these websites are appropriate for children. However, because of the nature of the Internet, it is impossible to guarantee that these sites will remain active indefinitely or that their contents will not be altered.

SHARE AND TAKE ACTION

Build a model tank.
Ask an adult to help you research craft ideas. Make sure your tank has tracks, armor, and a cannon.

Make armor.
Take two pieces of cardboard. Glue some foam squares between them. How do the foam and empty spaces add protection?

Compare tanks.
Ask an adult to help you look up different types of tanks. Which tank is stronger? Why?

GLOSSARY

accelerate To gain speed.

ammunition The objects (such as bullets and shells) that are shot from weapons.

armor A protective outer layer.

fuel A material, such as coal, oil, or gas, that is burned to produce heat or power.

hull The outer covering of a tank's top portion.

laser A device that produces a narrow and powerful beam of light.

obstacle An object to go around or over.

periscope A long tube with lenses and mirrors inside that is used to look over or around something.

target An object a weapon is aimed at.

trench A long, narrow ditch used a protection for soldiers.

INDEX

About the Author

Rachel Grack has been editing and writing children's books since 1999. She lives on a small ranch in Arizona. Rachel holds deep respect for the U.S. military. Her uncle served as a first lieutenant in the Marines. He was the pilot of an F-8 Crusader.